Analysis physical-mechanical of a triangular lattice

Juan Monsalve

Bibliographic information published by the German National Library:

The German National Library lists this publication in the National Bibliography; detailed bibliographic data are available on the Internet at http://dnb.dnb.de.

ISBN: 9783389068540
This book is also available as an ebook.

Trappentreustraße 1
80339 München

Print and binding: Books on Demand GmbH, Norderstedt, Germany
Printed on acid-free paper from responsible sources.

GRIN web shop: https://www.grin.com/document/1501808

Analysis physical-mechanical of a triangular lattice

Original Title: Análisis Físico-Mecánico de un retículo Triangular

Original version it is in Spanish language, this is a translate of the original document.

(translation was carried out Juan P. Monsalve G.)

Juan P. Monsalve G.

Abstract: This work offers a rigorous and detailed approach to an essential topic in mechanical and structural engineering, highlighting the importance of theoretical foundations in modern practice. It begins with a dedication to Stepan Tymoshenko, highlighting his engineering legacy, which adds valuable historical perspective. The use of triangular geometry is delved into, applying mathematical and physical rigor that guarantees informed conclusions. Additionally, it contains a writing style inspired by the 19th-century scientific texts, providing a distinctive touch that will appeal to readers interested in engineering history. This unique and meticulous approach makes the work a valuable resource for study and a significant contribution to the scientific literature.

Summary: This article is dedicated at memory of Stepan Prokofievich Timoshenko, known in the West-world as Stephen Timoshenko, who is widely recognized as the father of modern mechanical engineering. Born in Ukraine in 1878, he was a pioneer in the development of the theory of elasticity and resistance of materials, fundamental pillars in contemporary mechanical engineering and structural design. His work profoundly influenced the development of engineering as a scientific discipline, resulting in seminal texts such as "*Strength of Materials and Theory of Elasticity*", which remain key references in the education of engineers around the world.

Throughout his career, Timoshenko adopted a rigorous and mathematical approach, basing his analyzes on clear physical concepts and precise mathematical demonstrations. This same spirit is what we try to emulate with this article, which follows the writing and presentation style of the time in which Timoshenko made his most significant contributions. Therefore, we decided to take a retrospective approach, inspired by to the spirit of Mechanics and engineering originating from the beginning of the industrial revolution, which occurred between the 17th and 18th centuries.

This work aims to explain and substantiate each analysis with physical concepts and mathematical demonstrations, that are the interpretation of trigonometric relations and angular functions. We set aside subjective debates and return to the era of abstraction that Timoshenko embraced, focusing on the concepts of momentum and force distribution. The writing of this article is done in original typewriter handwriting and following a format for writing scientific writings from the late 19th century, a time of the rise of modern knowledge.

Clarification of the problem

Mechanical and structural engineers are familiar with triangular geometry, commonly used in trusses and other structures. There are standards and tabulators that help determine angles of inclination, positions and geometries, however, we rarely stop to reflect on the equations and principles that govern the physical-mechanical phenomena that prevail in this geometry that is so predominant in structural design.

This article combines the empirical experience accumulated in years of design with a rigorous theoretical-mathematical analysis, to study a triangular network. Firstly, we will present the case study followed by the developed free body diagram where the manifest forces will be exposed, taking into account the principle of symmetry to carry out the conceptual solution of an equilateral and scalene triangle. Next, a mathematical analysis will be carried out applying statics and Newton's third law to establish the general equations of the reactions and forces manifested in this geometry, taking asymmetry as a criterion, drawing conclusions in the process. Reach conclusions that provide a deeper understanding of the underlying mechanics of these structures. With this we reinforce the importance of returning to the theoretical foundations of engineering to address complex problems. The combination of empirical and theoretical analysis allows for a deeper understanding of the underlying mechanics in triangular structures, with practical applications in the design and optimization of structural components.

Timoshenko's legacy, which unifies theory with practice, remains to this day a guide for modern engineers to follow as we follow his example and that is why we seek to inspire future generations of engineers to explore the fundamentals scientists in their discipline with the same rigor and dedication that Timoshenko did at the time.

CASE STUDY (1) - EQUILATERAL TRIANGULAR LATTICE

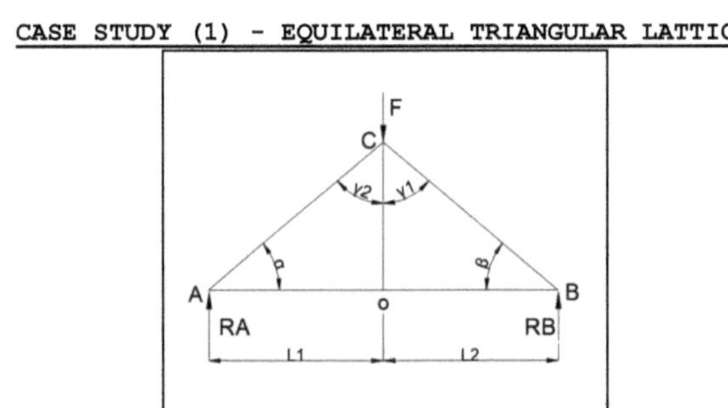

Figure 1
Source: Own elaboration

Figure 1 shows the D.C.L free body diagram of a hypothetical equilateral triangular lattice, its internal angles and the line of symmetry at points C and O.

Firstly, we establish that the triangle is Equilateral or Isosceles triangles, defined by

$$\alpha = \beta$$

We make a free body diagram at node C

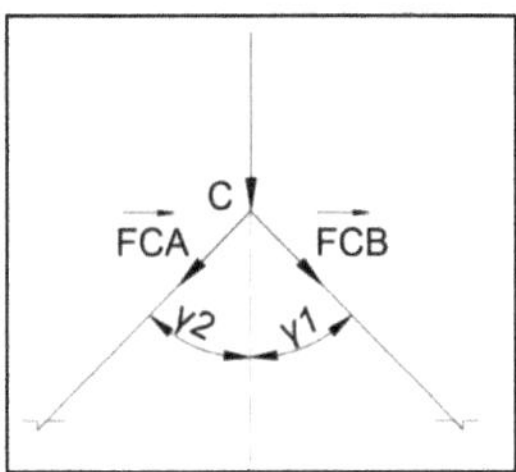

Figure 2
Source: Own elaboration

Figure 2 shows an enlarged view of Node C and the distribution of vector forces at said point.

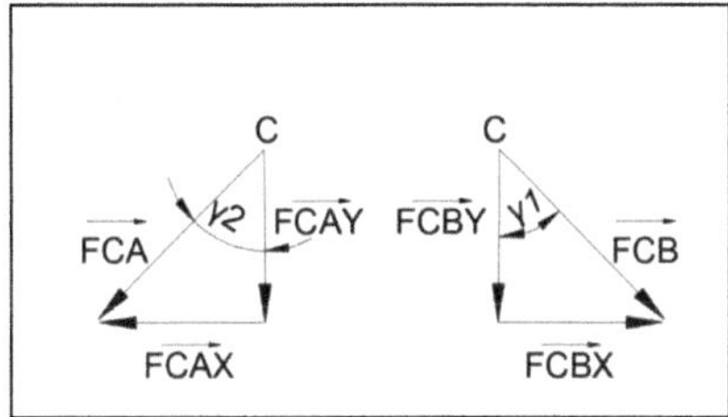

Figure 3
Source: Own elaboration

Figure 3 shows an enlarged view of Node C and the distribution of vector forces at said point after applying the principle of symmetry.

We know that the force $\vec{F}$ is going to be distributed in 2 superimposed angular forces in the sections $\overline{CA}$ and $\overline{CB}$ respectively, now the question is to know what percentage of force is manifested in each section? To solve this mystery, we will apply the principle of symmetry, which states:

In a body or element of symmetrical geometry and composition, physical phenomena are manifested in equal parts.

Based on this, we will divide the force $\vec{F}$ into two in the following way $\overrightarrow{F/2}$ + $\overrightarrow{F/2}$ and these forces will be equal to $\overrightarrow{FCAY}$ and (FCBY) $\overrightarrow{FCBY}$ respectively.

Now we know that angles $\gamma 1$ and $\gamma 2$ exist

Mathematically

$$\cos\gamma 2 = \frac{\|\overrightarrow{FCAY}\|}{\|\overrightarrow{FCA}\|} \Rightarrow \|\overrightarrow{FCA}\| = \frac{\|\overrightarrow{FCAY}\|}{\cos\gamma 2} \quad (1)$$

$$\cos\gamma 1 = \frac{\|\overrightarrow{FCBY}\|}{\|\overrightarrow{FCB}\|} \Rightarrow \|\overrightarrow{FCB}\| = \frac{\|\overrightarrow{FCBY}\|}{\cos\gamma 1} \quad (2)$$

Rewriting (1) and (2)

$$FCA = \frac{F/_2}{cos\gamma 2} \quad (3)$$

$$FCB = \frac{F/_2}{cos\gamma 1} \quad (4)$$

Being (3) and (4) the equations that govern the forces in the sections $\overline{CA}$ and $\overline{CB}$

$$\gamma 2 = 180° - 90° - \alpha = 90 - \alpha$$

$$\gamma 1 = 180° - 90° - \beta = 90 - \beta$$

$$\gamma 1 = \gamma 2$$

WE WIL PROCEED WITH THE STUDY IN NODE "A"

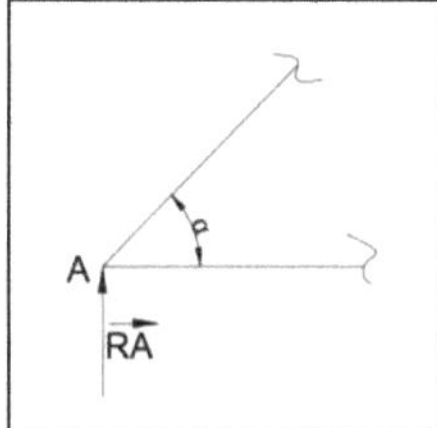

Figure 4
Source: Own elaboration

Figure 4 shows an enlarged view of Node A and the vector reaction at that point.

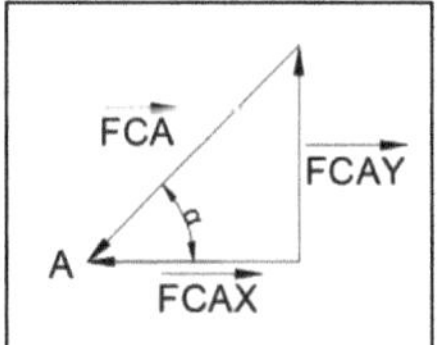

Figure 5
Source: Own elaboration

Figure 5 shows an enlarged view of Node A and the distribution of vector forces at said point.

In an analogous way to node "C", shown in Figure 2 and 3 and expressed in equations (1) and (2)

$$\sin\alpha = \frac{\|\overrightarrow{FCAY}\|}{\|\overrightarrow{FCA}\|} \Rightarrow \|\overrightarrow{FCAY}\| = \|\overrightarrow{FCA}\| * \sin\alpha = RA \quad (3)$$

WE WILL PROCEED WITH THE STUDY IN NODE B

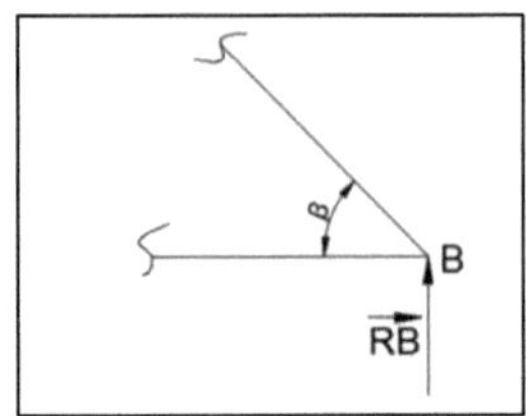

Figure 6
Source: Own elaboration

Figure 6 shows an enlarged view of Node B and the vector reaction at said point.

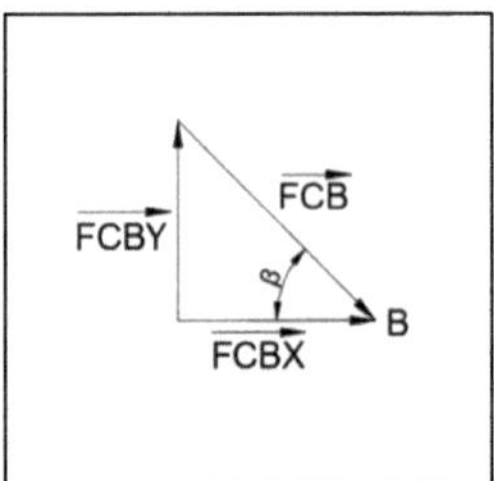

Figure 7
Source: Own elaboration

Figure 7 shows an enlarged view of Node B and the distribution of vector forces at said point.

Analogously to node "A"

$$\sin\beta = \frac{\|\overrightarrow{FCBY}\|}{\|\overrightarrow{FCB}\|} \Rightarrow \|\overrightarrow{FCBY}\| = \|\overrightarrow{FCB}\| * \sin\beta = RB \quad (4)$$

From figures 5 and 7 it can be deduced that $\overrightarrow{FCAY}$ and $\overrightarrow{FCBY}$ are equal to RA and RB respectively using Newton's 3rd law as a guarantee.

$$\text{Them: } RA = RB = {}^{F}/_{2} \quad (5)$$

It is understood that in a symmetrical triangle the reactions at the supports are equal to F/2, physically behaving like an arch.

CASE STUDY (2) - SCALANE TRIANGULAR LATTICE

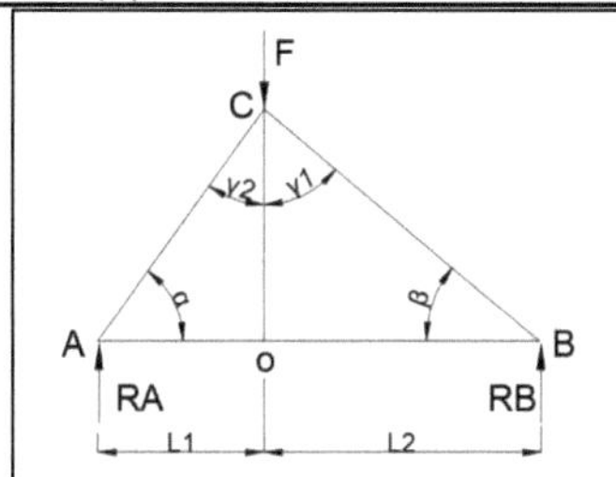

Figure 8
Source: Own elaboration

Figure 8 shows the F.B.D free body diagram of a hypothetical scalene triangular lattice, its internal angles and the line of asymmetry at points C and O.

For this case we will use the principle of Static Balance, which establishes that the sum of Moments and Forces in the structure must be equal to "0", (zero is read)

We set our reference point at "A"

Moment of $\vec{F}$ With respect to "A" $\overrightarrow{M_A}(\vec{F})=\overrightarrow{AC}X\vec{F}$

Mechanically the Moment is the vector product $\vec{R}X\vec{F}$

We hypothetically move $\vec{F}$ to "A"

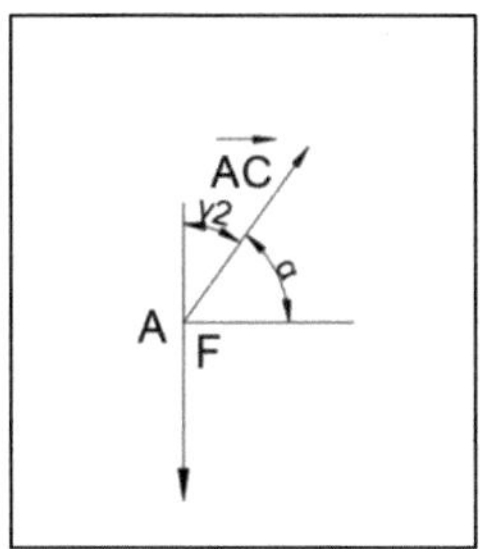

Figure 9
Source: Own elaboration

By Linear algebra it is proven and known that:

$$\| \vec{R} \| . \| \vec{F} \| . \sin(R \ngtr F)$$

where ≯ is refers to angle between R and F

Mathematical Proof

$$\| \vec{R} X \vec{F} \| = \| \vec{R} \| . \| \vec{F} \| . \sin\theta \qquad \text{(a)}$$

Firstly, it is defined

$$\vec{R} = (a_1 + a_2 + a_3)$$

$$\vec{F} = (b_1 + b_2 + b_3)$$

Module of vector

$$\| \vec{V} \| = \sqrt{V_x^2 + V_y^2 + V_z^2}$$

for analogy we establish

$$\| \vec{R} \| = \sqrt{a_1^2 + a_2^2 + a_3^2}$$

$$\| \vec{F} \| = \sqrt{b_1^2 + b_2^2 + b_3^2}$$

$$\| \vec{R} \|^2 = \left(\sqrt{a_1^2 + a_2^2 + a_3^2}\right)^2$$

$$\| \vec{F} \|^2 = \left(\sqrt{b_1^2 + b_2^2 + b_3^2}\right)^2$$

Them:

$$\| \vec{R} \|^2 = a_1^2 + a_2^2 + a_3^2 \ (i)$$

$$\| \vec{F} \|^2 = b_1^2 + b_2^2 + b_3^2 \ (ii)$$

now:

$$\| \vec{R} \|^2 . \| \vec{F} \|^2 = (a_1^2 + a_2^2 + a_3^2 \ ) . (b_1^2 + b_2^2 + b_3^2 \ )$$

$$= a_1^2b_1^2+a_1^2b_2^2+a_1^2b_3^2+a_2^2b_1^2+a_2^2b_2^2+a_2^2b_3^2+a_3^2b_1^2+a_3^2b_2^2+a_3^2b_3^2 \quad (iii)$$

We make vectorial scalar-product

$$\vec{R}.\vec{F} = (a_1 + a_2 + a_3).\ (b_1 + b_2 + b_3)$$

$$= a_1b_1 + a_2b_2 + a_3b_3$$

Them:

$$\left(\vec{R}.\vec{F}\right)^2 = (\,a_1b_1 + a_2b_2 + a_3b_3)^2 \quad (iv)$$

Solving the Vectorial Cross-Product

$$\vec{R}X\vec{F} = (a_1 + a_2 + a_3)X(b_1 + b_2 + b_3)$$

$$\begin{vmatrix} a_1 & a_2 & a_3 \\ b_1 & b_2 & b_3 \end{vmatrix} = \begin{vmatrix} a_2 & a_3 \\ b_2 & b_3 \end{vmatrix} - \begin{vmatrix} a_1 & a_3 \\ b_1 & b_3 \end{vmatrix} + \begin{vmatrix} a_1 & a_2 \\ b_1 & b_2 \end{vmatrix}$$

$$=(a_2b_3 - a_3b_2)-(a_1b_3 - a_3b_1) + (a_1b_2 - a_2b_1)$$

$$=(a_2b_3 - a_3b_2\,, a_3b_1 - a_1b_3\,,\ a_1b_2 - a_2b_1)$$

Now:

$$\| \vec{R}X\vec{F} \| = \sqrt{(a_2b_3 - a_3b_2\,)^2 + (\,a_3b_1 - a_1b_3)^2 + (a_1b_2 - a_2b_1)^2}$$

$$\left\|\vec{R}X\vec{F}\right\|^2 = (a_2b_3 - a_3b_2\,)^2 + (\,a_3b_1 - a_1b_3)^2 + (a_2b_1 - a_1b_2)^2$$

$$= a_2b_3{}^2 - 2a_2b_3a_3b_2 + a_3b_2{}^2 + a_3b_1{}^2 - 2\,a_3b_1a_1b_3 + a_1b_3{}^2 + a_2b_1{}^2 - 2a_2b_1a_1b_2 + a_1b_2{}^2 \quad (v)$$

solving (iii)-(iv)

$$a_1^2b_1^2+a_1^2b_2^2+a_1^2b_3^2+a_2^2b_1^2+a_2^2b_2^2+a_2^2b_3^2+a_3^2b_1^2+a_3^2b_2^2+a_3^2b_3^2-(\,a_1b_1 + a_2b_2 + a_3b_3)^2$$

Performing the perfect squared trinomial

$$(\,a_1b_1 + a_2b_2 + a_3b_3)^2 = a_1^2b_1^2 + a_2^2b_2^2 + a_3^2b_3^2 + 2\,a_1b_1a_2b_2 + 2\,a_1b_1a_3b_3 + 2a_2b_2a_3b_3$$

We have:

$$a_1^2b_1^2+a_1^2b_2^2+a_1^2b_3^2+a_2^2b_1^2+a_2^2b_2^2+a_2^2b_3^2+a_3^2b_1^2+a_3^2b_2^2+a_3^2b_3^2-a_1^2b_1^2 - a_2^2b_2^2 - a_3^2b_3^2 - 2\,a_1b_1a_2b_2 - 2\,a_1b_1a_3b_3 - 2a_2b_2a_3b_3$$

$$=a_1^2b_2^2 + a_1^2b_3^2 + a_2^2b_1^2 + a_2^2b_3^2 + a_3^2b_1^2 + a_3^2b_2^2 - 2\,a_1b_1a_2b_2 - 2\,a_1b_1a_3b_3 - 2a_2b_2a_3b_3$$
(vi)

Establishing (v) = (vi), this can be developed by the reader

Note (vi) = (iii)-(iv)

Therefore, we can establish the following equality.

$$\left\|\vec{R}X\vec{F}\right\|^2 = \parallel \vec{R} \parallel^2 . \parallel \vec{F} \parallel^2 - \left(\vec{R}.\vec{F}\right)^2$$

We know

$$\cos\theta = \frac{\vec{R}.\vec{F}}{\parallel \vec{R} \parallel . \parallel \vec{F} \parallel} \Longrightarrow \parallel \vec{R} \parallel . \parallel \vec{F} \parallel \cos\theta = \vec{R}.\vec{F}$$

Thus:

$$\left\|\vec{R}X\vec{F}\right\|^2 = \parallel \vec{R} \parallel^2 . \parallel \vec{F} \parallel^2 - \parallel \vec{R} \parallel^2 . \parallel \vec{F} \parallel^2 cos^2\theta$$

$$\left\|\vec{R}X\vec{F}\right\|^2 = \parallel \vec{R} \parallel^2 . \parallel \vec{F} \parallel^2 (1 - cos^2\theta)$$

By trigonometric identity

$$cos^2\theta + sin^2\theta = 1 \Rightarrow 1 - cos^2\theta = sin^2\theta$$

thus:

$$\left\|\vec{R}X\vec{F}\right\|^2 = \parallel \vec{R} \parallel^2 . \parallel \vec{F} \parallel^2 sin^2\theta$$

Simplifying

$$\parallel \vec{R}X\vec{F} \parallel = \parallel \vec{R} \parallel . \parallel \vec{F} \parallel \sin\theta \quad \text{(vii)}$$

It has been proven (a)

now:

$$R \ngtr F = 180 - \gamma 2 \Rightarrow R \ngtr F = (\pi - \gamma 2)$$

Trigonometrically $sin\,(180 - \phi) = \sin\phi$

It is demonstrated through the trigonometric circle of ratio 1 showed in figure 10

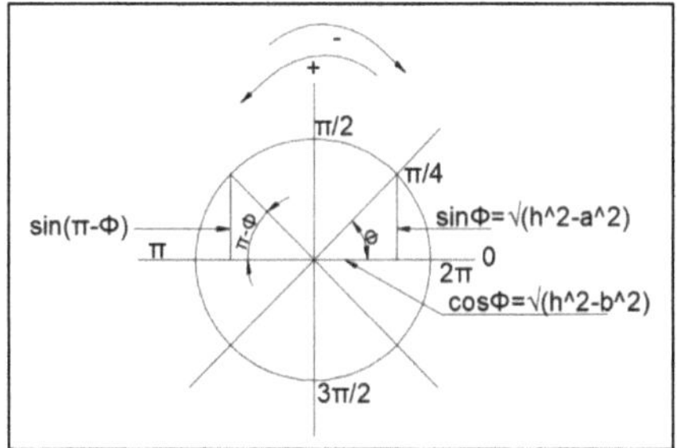

Figure 9
Source: Own elaboration

Figure 10 show the trigonometric circle of ratio 1

after:

$$\overrightarrow{M_A}(\vec{F})=\| \overrightarrow{AC} \|.\| \overrightarrow{F} \|.\sin(\gamma 2) \quad (6)$$

From figure 8 it is extracted

$$\sin(\gamma 2) = \frac{\|\overline{AO}\|}{\|\overline{AC}\|} \quad (7)$$

Substituting (7) in (6)

$$\overrightarrow{M_A}(\vec{F})=\| \overrightarrow{AC} \|.\| \overrightarrow{F} \|. \frac{\|\overline{AO}\|}{\|\overline{AC}\|} \quad (8)$$

$$\overrightarrow{M_A}(\vec{F})=F.\ \overline{AO} \quad (9)$$

Curious fact: in practice it is common to use the technique of "moving the force downward" or "Lowering the Force" in Spanish in the segment $\overline{AB}$ with origin in "O" to simplify the calculation of the sum of moments. Although this approach is mathematically correct, it is important to know that physically the force vector does not work that way. However, we are mathematically proven that this technique is valid.

summarizing moments respect to "A" demonstrated in the mathematical developed with formulas (6) and (9.

$$\sum M_A = -AO * F + (L1 + L2) * RB = 0$$

$$-L1 * F + (L1 + L2) * RB = 0$$

$$(L1 + L2) * RB = L1 * F$$

$$RB = \frac{L1}{(L1 + L2)} * F \quad (10)$$

how $(L1 + L2) = \overline{AB}$

$$RB = \frac{L1}{\overline{AB}} * F \quad (11)$$

Summarizing forces respect to the ordinates in a condition of equilibrium.

$$\sum F_Y = 0$$

$$-F + RA + RB = 0$$

$$-F + RA + \frac{L1}{\overline{AB}} * F \quad = 0$$

$$RA = F - \frac{L1}{\overline{AB}} * F$$

$$RA = F\left(1 - \frac{L1}{\overline{AB}}\right)$$

$$RA = F\left(\frac{\overline{AB} - L1}{\overline{AB}}\right) \quad (12)$$

like $\overline{AB} - L1 = \overline{OB}$

Rewriting (11)

$$RA = F\left(\frac{\overline{OB}}{\overline{AB}}\right) \quad (13)$$

From equations (11) y (13) is deduced:

If F applies at C and C is to the right of the midpoint of $\overline{AB}$

thus $RA < RB$

If F applies in C and C is to left of the midpoint of $\overline{AB}$

Thus $RA > RB$

If $\vec{F}$ applies at C and C is at the midpoint of $\overline{AB}$

Then RA=RB (Corroborating the principle of Symmetry).

CALCULATION OF THE FORCES MANIFESTED IN THE SECTIONS $\overline{CA}$ AND $\overline{CB}$

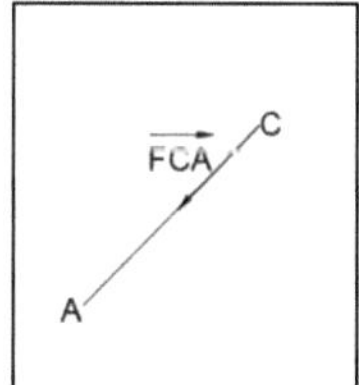

Figure 9
Source: Own elaboration.

Figure 11 shows the D.C.L free body diagram of the section $\overline{CA}$.

Applying the Principle of transitivity, we can move the Force $\overrightarrow{FCA}$ along its entire line of action at node A.

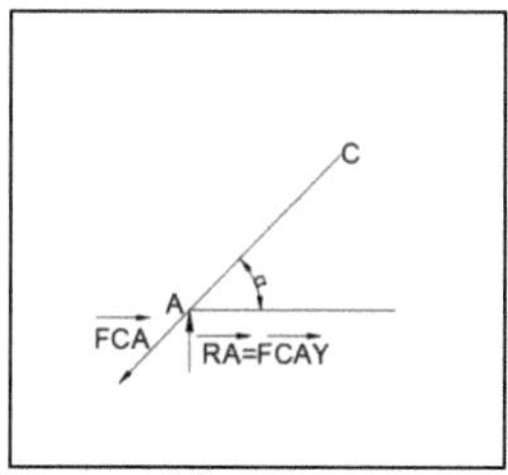

Figure 12
Source: Own elaboration.

Figure 12 shows the D.C.L free body diagram of the section $\overline{CA}$ And its vector reaction in A.

Note that the reaction in A $\overrightarrow{RA}$ is the component "Y" of $\overrightarrow{FCA}$ and it is equally to $\overrightarrow{FCAY}$.

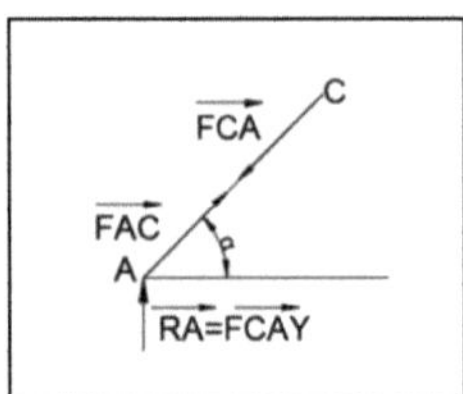

Figure 12
Source: Own elaboration.

Figure 13 shows the vector distribution of forces validated by Newton's 3rd law $\overrightarrow{FCA}$ *=*$\overrightarrow{FAC}$ *(Principle of action and reaction).*

Static balance must be maintained and therefore the forces manifested in section $\overline{CA}$ cancel each other.

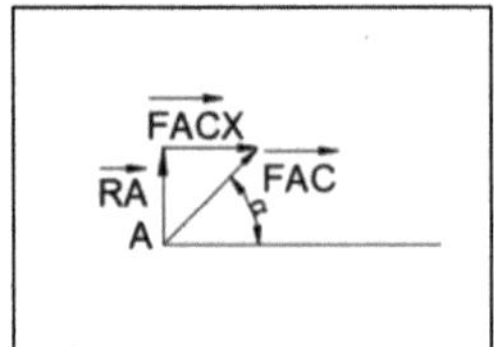

Figure 14
Source: Own elaboration.

Figure 14 shows that $\overrightarrow{RA}$ *=*$\overrightarrow{FACY}$*.*

now:

for trigonometric

$$\sin\alpha = \frac{\| \overrightarrow{RA} \|}{\| \overrightarrow{FAC} \|}$$

$$\overrightarrow{FAC} = \frac{\overrightarrow{RA}}{\sin\alpha} \quad (14)$$

$$\sin\alpha = \overline{CO}/\overline{CA} \quad (15)$$

substituting (13) and (15) in (14)

$$\overrightarrow{FCA} = \frac{\frac{F.\overline{OB}}{\overline{AB}}}{\frac{\overline{CO}}{\overline{CA}}} = \frac{\overline{CA} * F * \overline{OB}}{\overline{CO} * \overline{AB}} \quad (16)$$

In an analogous way it is done with $\overrightarrow{FCB}$ (It is left as an exercise for the reader)

$$\overrightarrow{FCB} = \frac{(\overline{AB} - L1) * F * L1}{\overline{CO} * L2} \quad (17)$$

In this way, we have established the formulas that determine the absolute value of the forces acting in the sections $\overrightarrow{CA}$ and $\overrightarrow{CB}$, as well as the equations that govern the reactions in equilateral triangle and scalene triangle configurations. Furthermore, we have shown that the practice of projecting the force applied at the upper vertex of the triangular lattice towards the span (AB) ¯ and multiplying it by the length of one of the lower vertices to calculate the moment is valid, since it is mathematically supported.

Finally

With this article there is significant potential to influence the design of more efficient and innovative structures. By deeply understanding how forces are distributed and manifested in a triangular lattice, it is possible to optimize geometry and material selection to create structures that are both light and robust.

For example, the use of triangular lattices with advanced materials could allow the development of structures that maximize material efficiency, minimizing weight without compromising structural integrity. This optimization not only results in a reduction in construction costs, but also improves the sustainability of projects by using fewer resources.

This approach can also be applied in earthquake engineering, where the ability of a structure to efficiently distribute loads is vital to resist seismic events. In summary, the conclusions of this work not only provide academic value, but also have a direct impact on engineering practice, allowing the design of more efficient, safe and sustainable structures.

RESULTS

General formulas for forces in a triangular lattice

Rewriting the lengths L1 and L2 in relation to the corresponding segments.

$$F\overrightarrow{CA} = \frac{\overline{CA} * F * \overline{OB}}{\overline{CO} * \overline{AB}}$$

$$F\overrightarrow{CB} = \frac{(\overline{AB} - \overline{OA}) * F * \overline{OA}}{\overline{CO} * \overline{OB}}$$

$$RA = F\left(\frac{\overline{OB}}{\overline{AB}}\right)$$

$$RB = \mathrm{F} * \frac{\overline{OA}}{\overline{AB}}$$

Example of application of the method.

A truss formed by triangular lattices is subjected to a distributed load of 4t. Calculate the forces on the inclined bars and the reactions.

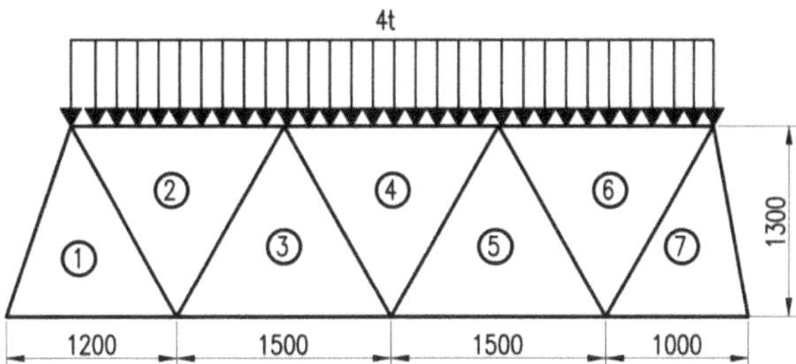

Figure 15
Source: Own elaboration.

Figure 15 shows a truss composed of triangular lattices.

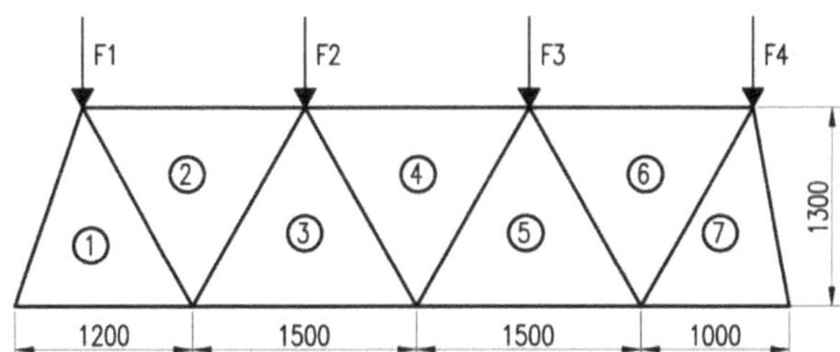

Figure 16 shows a truss composed of triangular lattices.

Figure 16 shows the discretization of the load into point forces at each upper node of each lattice.

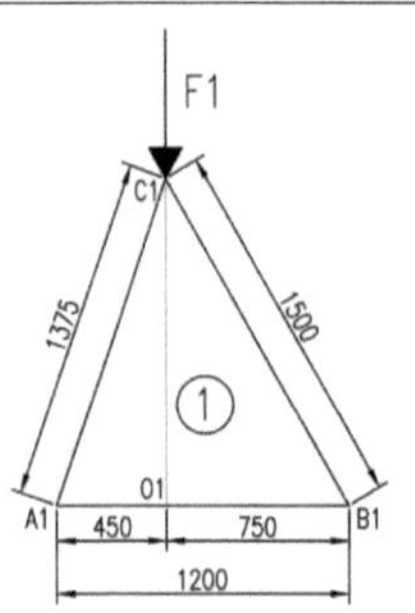

Figure 17, Reticle 1
Source: Own Elaboration

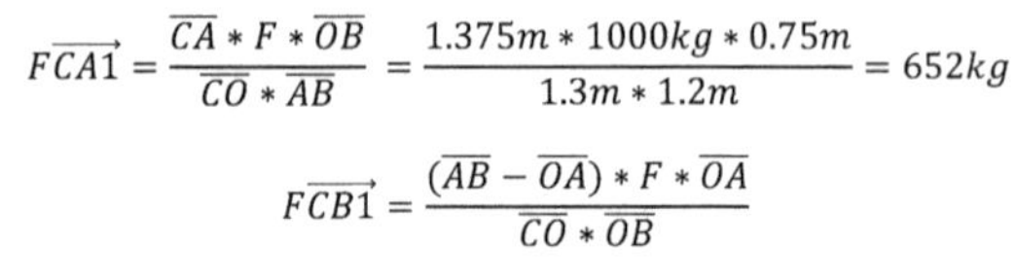

$$\overrightarrow{FCA1} = \frac{\overline{CA} * F * \overline{OB}}{\overline{CO} * \overline{AB}} = \frac{1.375m * 1000kg * 0.75m}{1.3m * 1.2m} = 652kg$$

$$\overrightarrow{FCB1} = \frac{(\overline{AB} - \overline{OA}) * F * \overline{OA}}{\overline{CO} * \overline{OB}}$$

$$\overrightarrow{FCB1} = \frac{(1.2m - 0.45m) * 1000kg * 0.45m}{1.3m * 0.75m} = 346kg$$

$$RA1 = F\left(\frac{\overline{OB}}{\overline{AB}}\right) = 1000\text{KG}\left(\frac{0.75\text{m}}{1.2m}\right) = 625\text{kg}$$

$$RB1 = \text{F} * \frac{\overline{OA}}{\overline{AB}} = 1000\text{kg} * \frac{0.45m}{1.2m} = 375kg$$

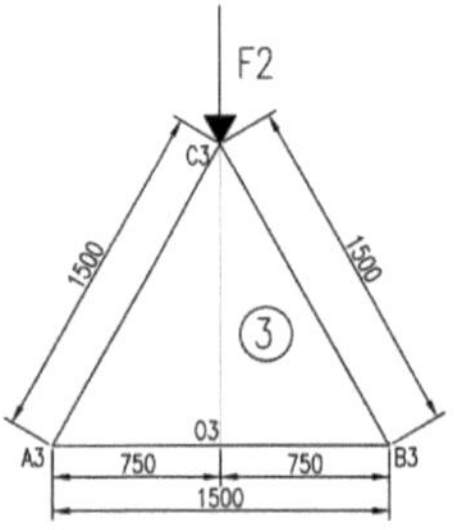

Figure 18, Reticle 3
Source: Own Elaboration

$$\overrightarrow{FCA3} = \frac{\overline{CA} * F * \overline{OB}}{\overline{CO} * \overline{AB}} = \frac{1.5m * 1000kg * 0.75m}{1.3m * 1.5m} = 578kg$$

$$\overrightarrow{FCB3} = \frac{(\overline{AB} - \overline{OA}) * F * \overline{OA}}{\overline{CO} * \overline{OB}}$$

$$\overrightarrow{FCB}3 = \frac{(1.5m - 0.75m) * 1000kg * 0.75m}{1.3m * 0.75m} = 578kg$$

$$RA3 = F\left(\frac{\overline{OB}}{\overline{AB}}\right) = 1000\text{KG}\left(\frac{0.75\text{m}}{1.5m}\right) = 500\text{kg}$$

$$RB3 = \text{F} * \frac{\overline{OA}}{\overline{AB}} = 1000\text{kg} * \frac{0.75m}{1.5m} = 500kg$$

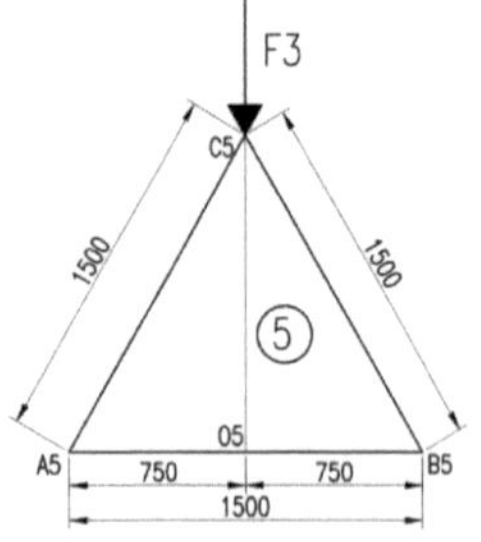

Figure 19, reticle 5
Source: Own Elaboration.

The distribution of forces in the reticle (5) is equal to the reticle (3)

$$\overrightarrow{FCA3} = \overrightarrow{FCA5}$$
$$\overrightarrow{FCB}3 = \overrightarrow{FCB}5$$
$$RA3 = RA5$$
$$RB3 = RB5$$

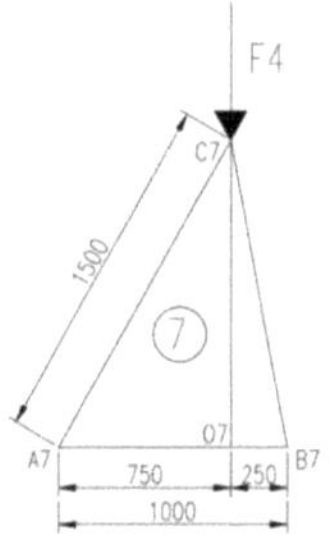

Figure 20, Reticle 7
Source: Own Elaboration.

$$F\overrightarrow{CA}7 = \frac{\overline{CA} * F * \overline{OB}}{\overline{CO} * \overline{AB}} = \frac{1.5m * 1000kg * 0.25m}{1.3m * 1m} = 288.5\ kg$$

$$F\overrightarrow{CB7} = \frac{(\overline{AB} - \overline{OA}) * F * \overline{OA}}{\overline{CO} * \overline{OB}}$$

$$F\overrightarrow{CB}7 = \frac{(1m - 0.75m) * 1000kg * 0.75m}{1.3m * 0.25m} = 578kg$$

$$RA7 = F\left(\frac{\overline{OB}}{\overline{AB}}\right) = 1000\text{KG}\left(\frac{0.25\text{m}}{1m}\right) = 250\text{kg}$$

$$RB7 = \text{F} * \frac{\overline{OA}}{\overline{AB}} = 1000\text{kg} * \frac{0.75m}{1m} = 750kg$$

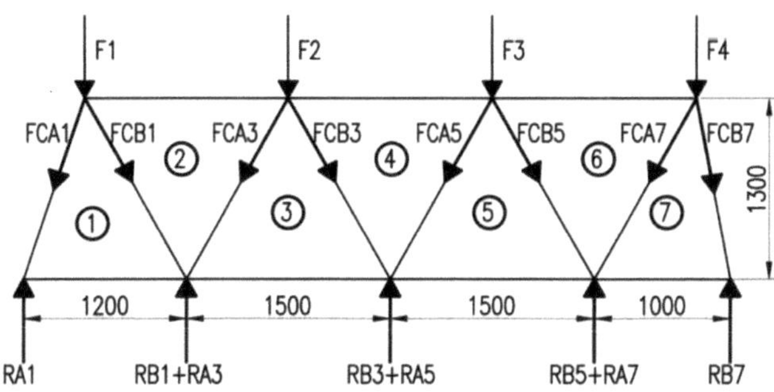

Figure 21
Source: Own Elaboration.

Figure 21 shows the vector distribution of forces in the armor.

CONCLUSIONS

This study on the physical-mechanical analysis of triangular lattices has achieved a greater understanding of the distribution of forces and reactions within configurations of equilateral and scalene triangles, establishing formulas and principles that can be applied in the design of efficient and resistant structures. The historical approach, inspired by the pioneers of modern science, has not only allowed a return to the theoretical foundations of engineering, but has also opened new ways to apply this knowledge in solving contemporary problems.

Beyond the results obtained, this work invites the academic and professional community to consider the integration of rigorous mathematical analysis with empirical practice, with the aim of exploring new applications in structural design and other areas of engineering.

We hope this will serve as a source of inspiration for future research, encouraging engineers and scientists to address current challenges with the same rigor and dedication that characterized the great masters of our discipline. With this, we not only reinforce the relevance of the theoretical principles, but also underline their potential to generate innovations that could influence the design and development of new technologies and structures in the future.

Referents

- Timoshenko, Stephen P. *Strength of Materials.* 3rd ed. Vol. 1. New York: D. Van Nostrand Company, 1955.

- Grossman, Stanley I. *Linear Algebra.* Philadelphia: Saunders College Publishing, 1988.

P.S: the references were made according to Chicago regulations